EXPLORING

WEIRD SCIENCE

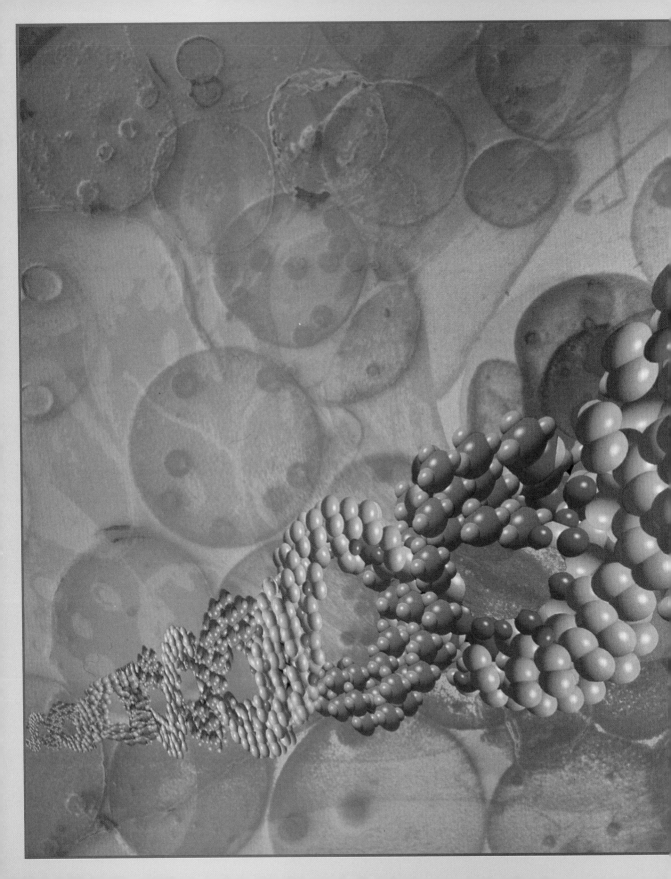

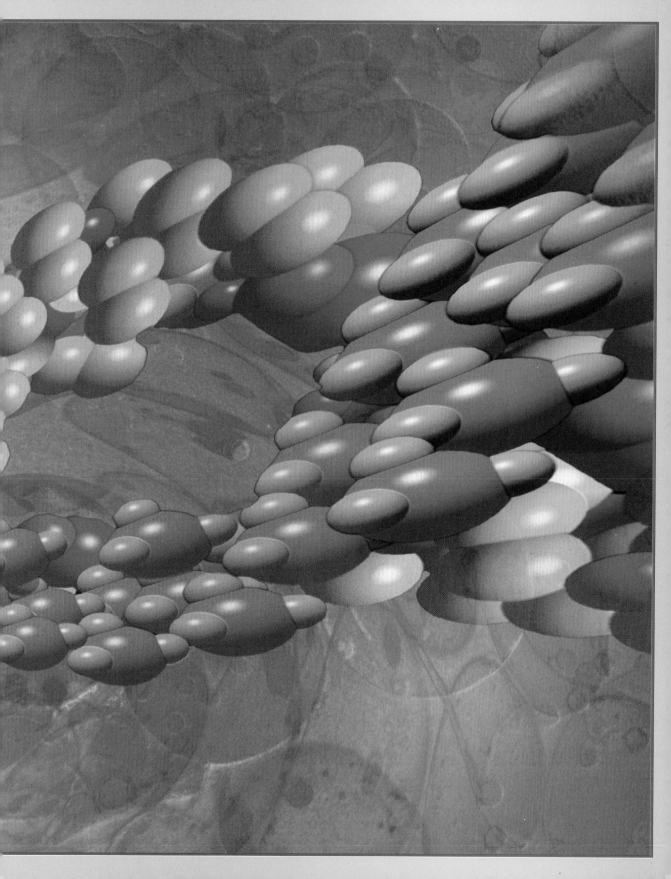

Scientific Consultant:
Malcolm Fenton, PhD.
The Dalton School

EYES ON ADVENTURE

EXPLORING

WEIRD SCIENCE

WRITTEN BY
REBECCA L. GRAMBO

kidsbooks®
Incorporated

WHAT AN IDEA!

Once upon a time, it would have been as weird to believe that Earth is round as it would be now to believe it is flat. In science, curiosity and the need to solve problems lead to investigation. Then new information and technology can turn strange ideas into reality.

NEW VIEW

In 1609, Galileo looked at the moon through a new invention, the telescope, and saw craters. That may not be a shock to you. But at that time, people believed that Earth was unique, and at the center of the universe. To them, the moon was a heavenly globe of light. But Galileo discovered that the moon is a world that, in some ways, resembles our own. Galileo changed our ideas about the universe forever.

AN INSPIRING BATH ▼

In the 3rd century B.C., Archimedes had a difficult problem to solve—he needed to measure the volume of the king's crown, an irregularly shaped object. As this story is often told, Archimedes solved his problem while taking a bath. Climbing in, Archimedes realized that he was pushing out an amount of water equal to his body's volume. "Eureka!" he cried, meaning "I have it!" All he had to do was put the crown in water and measure the water it displaced.

A SMALL WORLD

Plagues have killed millions throughout history. But before the mid-19th century, no one knew about germs. When Louis Pasteur proved his germ theory, he convinced doctors to boil their instruments and wash their hands. One of medicine's greatest discoveries, it led to a doubling of life expectancy and a population explosion.

This microscope image (shown in background) is the bacteria *E. coli*, which lives naturally in the human intestine and is necessary to a person's good health.

HORSE SENSE ▲

In 1872, two men made a bet about whether all four of a horse's hooves leave the ground when it gallops. To settle the bet, photographer Eadweard Muybridge set up 24 tripwires and cameras along a racetrack. As the horse galloped, it triggered the cameras one after the other. One photo showed all four feet in the air.

CHANGING ▶ TRUTH

Democritus (460-370 B.C.) said that all matter was composed of particles so tiny that nothing smaller was conceivable. He called them atoms, meaning "indivisible." Too weird for the time, the idea wasn't taken seriously for 2,000 years. After all, you can't see atoms—250 million of them lined up measure only one inch! Even smaller are the subatomic particles that make up an atom—protons, neutrons, and electrons.

IT'S RELATIVE ▲

Albert Einstein changed our view of time. His theory said that if one twin traveled through space while the other stayed on Earth, both twins would feel time pass normally. But time would slow down for the traveling twin, and he would age less than his twin on Earth. An experiment with an atomic clock carried by a jetliner proved Einstein correct. The clock showed less time passing than a clock on Earth.

7

THINK TANK

A frontier for discovery, the brain is far from being understood. How exactly do we think? To attempt to answer such questions, scientists are creating a kind of brain map, matching up various areas with specific functions.

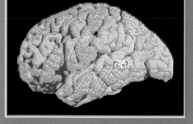

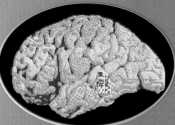

These scans show areas of the brain, in red, that are activated by words that are seen (top) and heard (left).

WAVE ACTION ▼

Electrical activity in the brain creates waves. In a procedure known as EEG, scientists attach electrodes to the skull and record these brain waves. EEG has been used to map the brain's functions, especially to study the role of sleep. Scientists have found that, like humans, other mammals and birds have periods of "dream sleep."

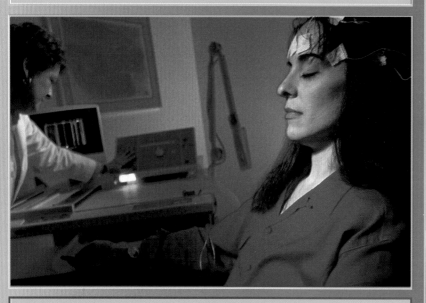

PROBING DEEPER

Scientists have inserted tiny probes, thinner than a human hair, to explore the brains of living people. The probes can detect the tiny electrical charges the brain uses to send messages. By sending electricity into certain parts of the brain, scientists can make people experience things that aren't really present, such as smells.

WONDER ORGAN

Everything we do is controlled by our brain, a mass of tissue weighing about three pounds. It's wired for communication by 1,000 billion neurons (nerve cells). These electronic signals carrying instructions, making thought and perception possible, can race along at 220 miles per hour inside your head!

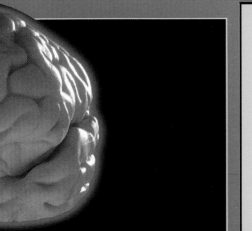

ATTRACTIVE MINDS

Did you know that your body has a magnetic field? This magnetic field, along with radio signals, makes it possible to take photographs of your brain. This kind of photography is called Magnetic Resonance Imaging (MRI). Hoping to improve MRI, one scientist has created detailed maps of magnetic fields in humans.

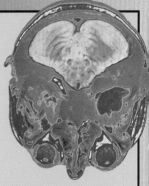

▲This MRI scan shows a section of the brain and eyes.

▲ RIGHT AND WRONG

In 1848, an explosion sent a metal rod through the head of Phineas Gage. Amazingly, he survived, but his personality changed. Previously a calm, hardworking man, Gage became grouchy and irresponsible, used profane language, and lied to friends. Using computer imaging, scientists are studying this case and others to pinpoint a possible "moral" center in the brain—a place that helps a person to tell right from wrong.

MIND TO BODY

When you look at your left arm, how do you know that it's yours? On each side of the upper brain, there is a patch of neurons called the parietal cortex. This area is involved in sensing movements and objects on the opposite side of the body. People who suffer injuries, such as a stroke, to the right parietal cortex may deny that their left arm or leg is their own. One patient claimed that her arm belonged to her mother, who had left it behind in the hospital bed.

CURIOUS CURES

Until about 100 years ago, bloodletting was a normal treatment for many illnesses, including fevers and pneumonia. The doctor would cut the patient in a particular area of the body and let a certain amount of blood flow out. Now we think of this as weird, because loss of blood is harmful, not helpful. But some of the treatments we use today seem every bit as strange.

◀IN THE DIRT

Where do you find wonder drugs? Scientists have poked around in the mud and muck to find them. The organism that produces streptomycin, an important antibiotic, was first discovered on a hen. Scientists then searched the whole farm and found the organism in the heavily manured soil around the henhouse.

OUCH! ▲

It may look painful, but acupuncture is a Chinese medical procedure that actually helps relieve pain. Even pets get relief. The physician may insert needles at more than 360 points in the body, and then twirl the needles or send electrical currents through them. In China, acupuncture is often used along with drugs during brain surgery.

◀GROSS WORKS

Some ancient remedies are still used today. After reattaching fingers or toes, doctors may apply leeches—blood-sucking worms— to help keep tiny blood vessels from clogging. Maggots are used to remove dead flesh. And physicians sometimes pack deep wounds with the same thing used by Egyptians 4,000 years ago—sugar.

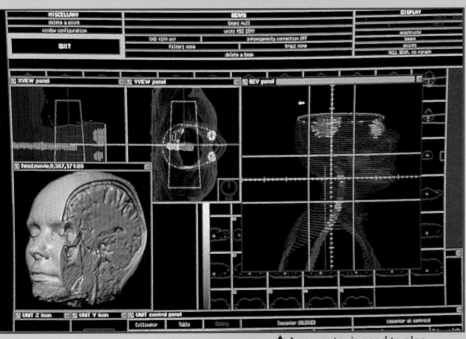

KILL TO HEAL

Although X rays were first observed by accident, scientists have learned much about radiation and put it to good use. Too much radiation can be fatal to humans, but, by focusing those killing powers on diseased cells, radiation can be used to treat cancer.

▲ A computer is used to plan radiation treatment.

GROWING CURES

We often think of plant remedies as something used only by ancient people, but almost 80 percent of people today rely mainly on this traditional medicine. Nearly 50 percent of medicines on the market come from plants. More than 25 percent of all medicines come from ingredients found in rainforest plants and animals.

◀ Marie Curie (1867-1934), the only person to win two Nobel Prizes in science, was the first to describe the process of radiation.

FROG MEDICINE

Long used by Ecuadorian Choco Indians as a weapon, the venom from the poison arrow frog can help as well as hurt. One compound in the venom acts as a painkiller that is 200 times better at fighting pain than morphine, a drug used in hospitals.

THE GENE SCENE

Our genes carry DNA (deoxyribonucleic acid)—the substance that makes us unique individuals. Although it fits into a single cell, the human DNA molecule is about six feet long. If you stretched out all the DNA molecules in a baby, and put them end to end, they would reach 114 billion miles—thirty times the distance between the Sun and Pluto!

SEEING DOUBLE ▲

In 1996, Dolly, a sheep in Scotland, became the first mammal ever cloned from adult cells. A clone is a genetic duplicate—an animal or plant that is identical to another. Scientists replace the DNA of an egg cell with the DNA from another cell. The egg then develops into a clone of the animal that provided the new genetic material.

◄ DNA DETECTIVES

Who did it? DNA can link criminals to a crime scene better than any fingerprint. Scientists use blood or skin cells found at the scene to create a genetic profile, which they compare to the suspect's. In one case, the DNA taken from cat hairs found at the scene of a crime was matched to that of the suspect's pet. The man was found guilty and convicted.

FAMILY TREES ▼

By using DNA, it's possible to trace family relationships. Scientists studying the DNA of a 9,000-year-old skeleton from Cheddar, England, discovered that he had a living descendant—the local school teacher! That's a family tree with long roots.

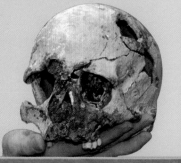

▲ ZONKEY!

What happens when you cross a zebra with a donkey? You get a crossbreed, an offspring with a mix of genes from both parents. Through biotechnology, scientists may isolate a specific gene from one species and give it to another, creating new varieties of plants and animals.

GENE ALARM

Do you wake up at the same time every day? We all have a biological clock that controls our sleep cycle. Researchers have discovered that, in mice, a gene controls this clock.

SAVING SPECIES

Test-tube tigers? Scientists are using frozen sperm and eggs to create baby animals that belong to endangered species. By fertilizing the egg in a test tube, it is possible to breed animals separated by great distances that might not otherwise mate.

▶ DINOS TODAY?

Bad news for *Jurassic Park* fans. In the movie, scientists used DNA taken from blood in the stomachs of biting insects trapped in amber (fossilized tree resin) to re-create dinosaurs. In reality, this would be impossible. However, scientists have discovered some pieces of dinosaur DNA in fossil bones.

MICRO WORLD

With powerful instruments, you can enter the world of the very small. An *optical* microscope gets you into the main parts of a cell, magnifying things up to 1,500 times. An *electron* microscope makes things more than 500,000 times life size. A *scanning tunneling* microscope, a type of electron microscope, can magnify up to one million times—bringing the structure of an atom into focus!

◀ In this photograph, blood cells are shown moving through the smallest branch of an artery.

▲Pollen can cause allergic reactions known as hayfever.

MICRO WARFARE
Viruses represent one of the biggest challenges to overcoming infectious diseases. When a virus, such as the common cold, invades a cell, it forces the cell to make copies of the virus. To attack the virus is to attack the body's own living cells. No cures exist, only preventive vaccines.

▲This weird-looking virus is responsible for causing warts.

AH-CHOO!
The body has a remarkable defense against infection. But in some people, this immune system gets a little confused. It recognizes certain harmless particles as dangerous and tries to fight them off. An allergic reaction is the result, which can be minor, like a sneeze, or, in extreme cases, life-threatening.

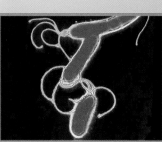

TOUGH GUYS
You think you're tough? Some bacteria live in boiling hot water. Others survive radiation that would kill a human. Scientists have even found bacteria living 4,500 feet below the ground. These bacteria appear to survive in total darkness on nutrients they extract from the rocks.

MITE-Y SMALL
You never see them, but the tiny spider-relatives known as mites (left) are all around us, even in the cleanest house. They crawl through carpet, prowl your bed, and even live in your hair! But don't confuse them with the blood-sucking, foul-smelling insects known as bedbugs (above). Mites actually help keep your house clean by feeding on flakes of dead skin.

◀BEACH BABIES
It's just a beach, or is it? Living in the sand is one of the richest animal communities found on Earth. These tiny meiofauna (MY-oh-faw-nah) are strange creatures—some have heads covered with whirling hairs, others cling to sand grains with hooked claws. A single handful of wet sand may hold 10,000 of these animals.

NATURAL ODDITIES

Imagine finding a new life form, or talking to animals. In nature, scientists encounter some pretty strange things, and engineer some weird ways of getting inside an animal's world.

▲DOWN DEEP

In the murky waters of the Amazon River, where light penetrates to only a few inches below the surface, many new species of fish are being discovered. Mostly blind, they rely on electrical signals to navigate the dark river water.

WHITE WONDER▼

Coloration exists for a reason in nature, often as camouflage against predators. That's why it's strange to find an animal colored differently than the norm—like this pure white emperor penguin spotted in Antarctica. Sometimes white animals are albinos, which lack pigment and have red eyes. However, this penguin is not an albino, but a kind of genetic mutant rarely seen.

◀ SPYING CRABS

How do you know if an animal is using its eyes, and not its sense of smell? Some scientists mounted a tiny spy camera above the eyes of a male horseshoe crab to make a film of what the crab saw—female crabs. Meanwhile, they monitored electrical responses in the nerves that connect the eyes to the brain. Back in the lab, male crabs who saw the film responded in much the same way as the crab on the loose, proving that the eyes were at work.

PUPPET MOM ▶

Raising animals in captivity leads to all sorts of trouble. The animals see humans as protective parents and never learn to survive on their own. One solution to raising bald eagles and other birds is to use puppets that look like the adult animals. The babies receive food from a beak, not a human hand.

SMART TALK ▲

Have you ever wondered if animals have their own language? A study on prairie dogs used a computer to correlate squeaks and chirps with events happening at the time. Results suggested that prairie-dog talk distinguishes a coyote from a German shepherd and a man from a woman.

PLANT FIGHT

When disease strikes, plants fight back. At least, that's what some scientists think. Plants produce the chemical salicylic acid, or aspirin, when they are sick. This self-made medicine boosts a plant's immune system. It also stimulates other plants to be on the defense. In the future, instead of spraying toxic pesticides, maybe people will treat plants with aspirin.

▼ARTY APES

Believe it or not, people can communicate with animals. The famous gorilla named Koko began learning sign language in the 1970s. She uses more than 1,000 signs to relay her thoughts. One of her favorite activities is painting. When asked what her blue, red, and yellow painting was, Koko answered, "bird."

LOST IN TIME

When the first fossils were found, no one knew what they were. Fossils are traces of plants and animals preserved in the earth. Today, scientists called paleontologists (pay-lee-on-TALL-oh-jists) are unraveling the stories that fossils tell.

STRANGE CREATURES

Preserved in 530-million-year-old rock in Canada, called the Burgess Shale, are some of the weirdest animals that ever lived on Earth. One was a worm with five eyes and a trunk.

FEATHERED FINDS ▲

A rich fossil bed discovered in China in the 1990s is remarkable because of the amount of detail preserved in the rock. Among its treasures are lots of bird fossils showing feathers, including the oldest beaked bird.

HUGE INHABITANTS ▼

Fossils reveal that South America had some hefty inhabitants—an eight-ton, meat-eating dinosaur we call Giganotosaurus, and the plant-eater Argentinosaurus (below), which weighed in at around 100 tons. Other discoveries include two-foot spiders and a forty-foot-long crocodile.

ANCIENT FOREST

A subtropical forest only 680 miles from the North Pole? On Axel Heiberg Island, scientists found the *mummified* remains of one. The forest grew 45 million years ago when the area was warm and swampy. The really weird thing is that the wood could still burn today.

UP IN THE AIR ▶

If you had been in Texas about 65 million years ago, you might have seen something impressive overhead—the largest flying creature ever, a reptile called Quetzalcoatlus (KWET-zal-KWA-tel-us). It weighed about 190 pounds and had a wingspan of about 40 feet—the size of a small plane.

▼ CYBER DINOS

If you want to see how dinosaurs moved, you have to bring them to life. Some scientists have done just that—in cyber space. They used software to reconstruct sauropods, such as Apatosaurus (also known as Brontosaurus). This 100-ton dino may have broken the sound barrier, creating a sonic boom by using its 3,500-pound tail like a whip!

THE GREAT DYING

The dinosaurs died out 65 million years ago, but about 245 million years ago, a much larger extinction event took place. During what is known as the Great Dying, as much as 96 percent of all plant and animal species on Earth may have been wiped out.

COLORFUL DINO?

Ever wonder what color Stegosaurus really was? Researchers looking at fossils have identified structures and pigments responsible for colors. Using this information, they've determined that a 370-million-year-old fish was dark red on top and silver underneath. Could the dinosaurs be next?

DIGGING HISTORY

Scientists are still trying to piece together human history, especially the prehistoric period when no written language existed and few records were kept. Archaeologists and anthropologists are the detectives looking for the clues.

OLD ART ▲

The last time human eyes saw these cave paintings, discovered in 1994 in the Ardeche River canyon of France, may have been 30,000 years ago. Over 300 paintings represent Ice Age animals that include horses, bears, rhinoceroses, and bison.

▼ MODEL PAST

Starting with a skull and using charts that show the thickness of tissue, scientists can re-create faces from long ago. They build models of the face from clay or use computers to make a three-dimensional image.

JIGSAW HEAD ▲

Bones and artifacts, discovered on a dig or purely by chance, can take scientists back millions of years. Richard Leakey and his team of researchers found 150 skull fragments and other bones in Kenya dating almost 2 million years ago. The team reconstructed the skull, and named it "handy man" because it was found alongside stone tools.

◄ PEKING MAN

A famous collection of fossils known as Peking Man was discovered in 1921 in China. Thought to be 400,000 years old, the bones were the first of their kind found in Asia. Stone tools were also found in the cave.

VIKING ◀ VISITS

Did some European explorer beat Columbus to the Americas? The remains of eight buildings and other artifacts discovered in Newfoundland, Canada, in 1960 finally confirmed that Vikings got here long before 1492. But their attempt to settle the continent failed.

ICE MAN

Imagine hiking in the Alps and finding a body that's over 5,000 years old, preserved like a time capsule in a glacier. That's what happened in 1991, when two mountaineers discovered the "Ice Man."

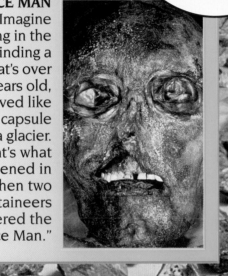

BOG BODIES

In cool places where water stands still, layers of dead plants pile up sometimes 40 feet thick underneath living plants, and a bog is formed. Called peat, the dead plant matter is useful as a kind of fuel. People digging for peat have made some bizarre discoveries. They have found perfectly preserved bodies—whiskers and clothes included—dating from over 2,000 years ago.

HANDLE WITH CARE

Scientists can use high-tech methods to learn all about a mummy without damaging it. Using Computer Assisted Tomography (CAT) scanning, they take a series of X-ray pictures of the mummy still in its wooden case. Then the computer adds up the images to create a 3-D image.

SPOOKY SPACE

In the vast reaches of space, almost everything, like the death of a star, seems weird and mysterious. At the end of its life, a star may collapse in upon itself and form a black hole. Sounds interesting, but you don't want to visit. A black hole's gravity is so strong that it pulls in everything around it, even light.

Gas exhaust from this dying star moves in twin jets at a velocity greater than 200 miles per second.

SPACE INVADERS

Earth gets pelted by about 20 tons of rock a day from meteorites, asteroids, and comets. Most of the material arrives as grain-sized particles. Really big impacts that cause widespread destruction occur only about once every 100 million years.

COSMIC SNOWBALLS ▲

Some scientists say that up to 30 house-sized chunks of ice are hitting our planet's atmosphere every minute. Each of these big snowballs weighs about 20 tons. We're not up to our noses in snow because the ice melts while still thousands of miles above Earth.

▶ BIG LIGHT

The Pistol Star cannot be seen without a telescope, yet it sends out 10 million times more energy than our Sun! Newly discovered, it is believed to be 186 million to 280 million miles across—about as big as Earth's orbit around the Sun.

ANOTHER COMPANION

Earth travels with more than just a moon—Asteroid 3753, just over three miles in diameter, is trapped by Earth's gravity into a strange orbit between Mercury and Mars. It takes 770 years for this asteroid to orbit Earth.

◀ HOT BLOB

When electrically charged gas gets ejected from the Sun, a blob of it may weigh tens of billions of tons, travel about 2.2 million miles per hour, and carry enough energy to boil off the Mediterranean Sea. When it hits Earth's atmosphere, a geomagnetic storm occurs, sending curtains of colorful light called auroras over the polar skies.

◀ HOME UNKNOWN

What is the best-mapped planet in our solar system? Not Earth! More than 98 percent of Venus's surface has been mapped, thanks to the explorer spacecraft *Magellan*. Here on Earth, more than two-thirds of the surface is covered with water, and much of it cannot be charted as accurately as the Venusian surface.

EERIE EARTH

Our home planet harbors a store of weird happenings, from powerful earthquakes to mysterious, phenomenal events. In 1811, things were rocking in New Madrid, Missouri. The area experienced the strongest earthquakes in recorded American history. More than 3,000 square miles of land were visibly damaged and, for a brief time, the Mississippi River flowed backwards!

SPIN CITY

The Moon's gravitation acts like a brake on Earth, slowing down its rotation. Each year our days get about 20 millionths of a second longer. But something else is having the reverse effect. Scientists have discovered that dams, in concentrating water away from the equator, are making the planet spin faster!

COMPASS CHECK ▲

The motion of molten metal in Earth's outer core produces an electric current, causing the planet to behave like a giant magnet. Earth's magnetic field extends about 37,000 miles into space and protects us from some of the Sun's most harmful particles. Curiously, the magnetic field reverses from time to time—the last occasion was about 700,000 years ago.

HIDDEN LAKE

A freshwater lake about the size of the state of New Jersey lies under more than two miles of ice in eastern Antarctica. The lake may be a mile deep in places. Bacteria and other life forms in Lake Vostok haven't been in contact with the surface for at least 50,000 years, and perhaps as long as three million years.

200 Million Years Ago

100 Million Years Ago

Today

▲ON THE MOVE

Think you can stand perfectly still? Guess again. Not only is Earth rocketing around the Sun at 66,600 miles per hour; it's spinning on its axis at about 1,000 miles per hour. Even the ground beneath you is never motionless, as big slabs of Earth's crust, called tectonic plates, are moving very slowly.

◀ Scientists believe that the continents were once all joined together, and that the movement of Earth's crust caused them to break apart. The Americas continue to drift away from Europe and Africa at a rate of one inch per year.

BIG BANG

When the Indonesian volcanic island of Krakatau blew itself to bits in 1883, a sound like distant cannons reached places nearly 3,000 miles away! The blast caused huge waves, called tsunamis (sue-NAH-mees), as far away as South America and shot ash an estimated 50 miles into the air.

PINGO PONG

In Siberia and northern Canada, frost stays in the ground year round, reaching as deep as 1,650 feet. Houses have to be built on stilts so they don't melt the frost and sink into the mud. Sometimes the frost pushes up piles of earth as high as 230 feet to create mounds called pingos.

DREAMS TO DEVICES

Scientists have built some weird and amazing things. Researchers in California have created the most powerful magnet in the world, 250,000 times as strong as Earth's magnetic field. Other scientists have created a light brighter than every star in our galaxy, concentrated in a spot the size of a pinhead. Called the Vulcan laser, it may enable scientists to look into living cells and capture molecules in action.

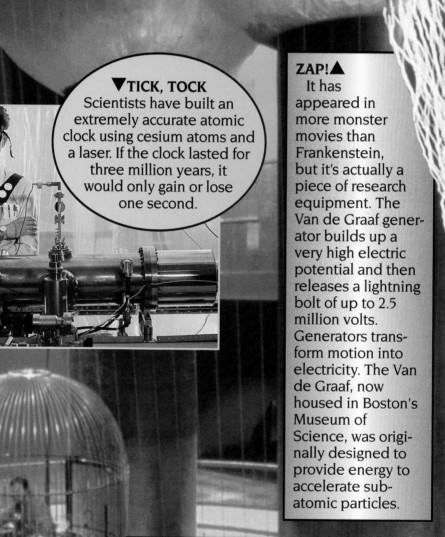

▼TICK, TOCK
Scientists have built an extremely accurate atomic clock using cesium atoms and a laser. If the clock lasted for three million years, it would only gain or lose one second.

ZAP!▲
It has appeared in more monster movies than Frankenstein, but it's actually a piece of research equipment. The Van de Graaf generator builds up a very high electric potential and then releases a lightning bolt of up to 2.5 million volts. Generators transform motion into electricity. The Van de Graaf, now housed in Boston's Museum of Science, was originally designed to provide energy to accelerate sub-atomic particles.

MOTOR MOTH
Accepted ideas about flight say insects should not be able to get off the ground. Yet insects are some of the most agile fliers in the world. To help solve the mystery, scientists have built a giant moth to study. With motors and gears instead of muscles, the moth flaps its three-foot wings in the same way as a real one.

Using an ▲ electron beam, nanotechnology could record 29 volumes of the *Encyclopedia Britannica* on this tiny target made of aluminum fluoride.

TEENSY TECHNOLOGY

The smallest science of all is nanotechnology— building things the size of molecules. This gear (below) is smaller in diameter than a human hair and 100 times thinner than a sheet of paper. Etched into the surface of a silicon wafer, a "micro- motor" could one day be used in medicine—allow- ing miniature robots to roam your bloodstream and heal injuries.

ROBO PALS ▲

Although robots came straight from science fiction, scientists are putting them to work in the real world. Robotics is the sci- ence of giving machines certain capabilities that are usually found only in living creatures. The roboticist (ROE-bah-tih- cist) above has built robots that can find land mines, and solar machines that compete for sunshine.

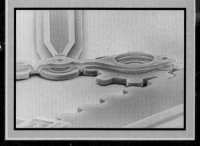

◄ TAKE ME HOME

Every driver would like to avoid traffic jams. A new comput- erized guidance system takes information from the city's traffic computers and sends it to displays in cars. The system figures out the quickest route but ignores any that show traffic jams.

STRANGE STUFF

Sometimes when we think of science, images of smoking beakers come to mind. In the laboratory, scientists can discover or create some pretty weird things.

▼ Solid carbon dioxide, known as dry ice, becomes a gas when mixed with water, producing thick clouds of fog.

◀ BUCKYBALLS?

◀ BUCKYBALLS?
Sixty carbon atoms arranged in a perfect sphere make up a molecule of buckminsterfullerene, or a "buckyball." Because of their shape, these molecules have been named after Buckminster Fuller, an engineer who developed the domed stadium. Buckyballs can be found in the soot floating around after you blow out a candle. Other such hollow carbon molecules are called bunny balls, buckybabies, and bucky onions.

WHAT'S THE MATTER?
Solid, liquid, gas, plasma—the four states of matter. Now there's a weird fifth form—clusters. Clusters are groups of atoms that seem to be somewhere between atoms and the regular-sized world. They have strange electrical, optical, and magnetic properties.

STRETCHED THICK
If you punch a pillow, your fist leaves a dent in the surface. But some strange substances called auxetic (og-ZED-ik) materials don't act this way. When stretched, they get thicker. A pillow made of auxetic material would expand when punched.

◀ LIGHTWEIGHT
Scientists have made silica, the raw material for glass, into something new—an aerogel (AIR-oh-jell). The substance contains as much as 96 percent air and weighs far less than glass. Because it's so clear, aerogel is hard to find once it is placed on a lab bench. Here, it is shown supporting a weight and a penny.

GENIE IN A BOTTLE?

No, it's not magic. It's a hologram, a photograph taken with laser light. One set of light waves is sent to the object, which reflects the light onto film. Another set of light waves, sent directly to the film, intersects the reflected light waves and creates a 3-D image. You can walk all the way around the picture to view it from different angles.

VIRTUAL LAB

Sometimes it's difficult, or completely impossible, for scientists to conduct their studies in real-life places. Virtual reality, simulated by computers, takes them there—or appears to.

◀ In a virtual molecular interaction, a researcher can pick up molecules, reorient them, and study what occurs.

29

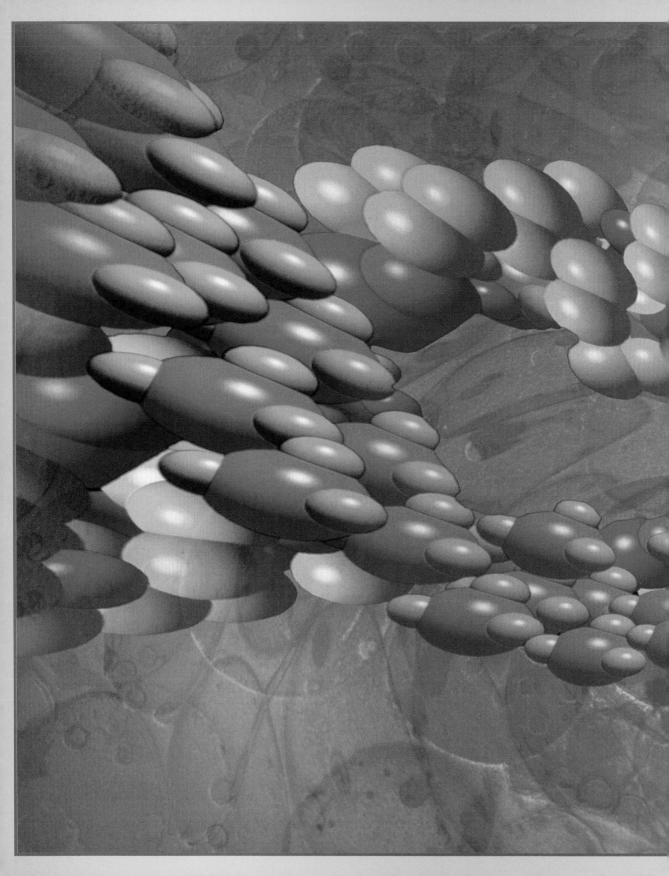

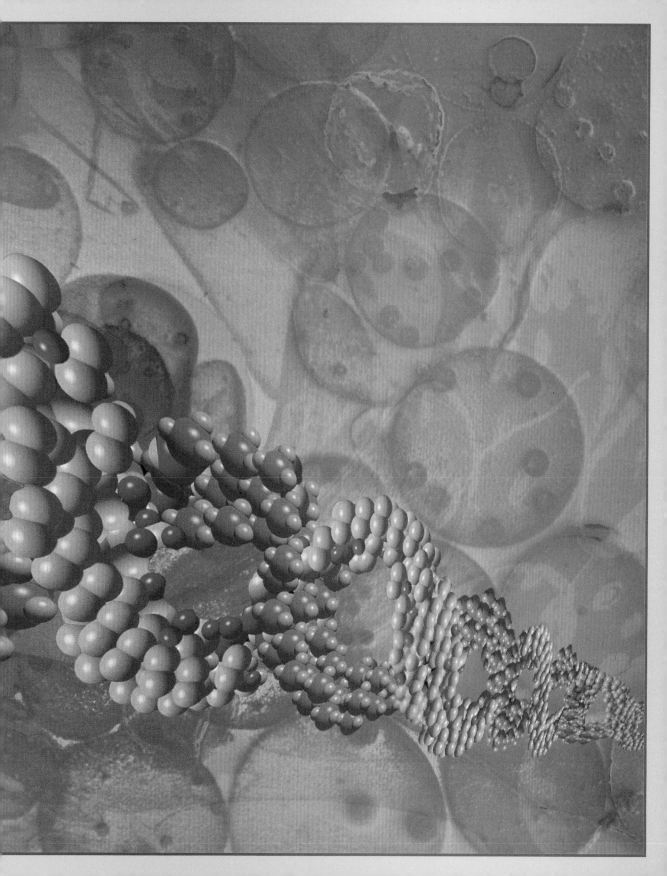